IMAGES
of America

THE NAVY IN SAN DIEGO

The nuclear-powered aircraft carrier *Ronald Reagan* returns to port via the San Diego channel in 2006 with her crew manning the rail. (U.S. Navy.)

ON THE COVER: A navy recruit stands at the temporary naval training station during the summer of 1918. In the four years of its existence, nearly 4,000 sailors passed through this temporary boot camp housed in the vacant buildings of the Panama-California Exposition in Balboa Park before the opening of the naval training station in Loma Portal in 1923. (Maritime Museum of San Diego.)

Bruce Linder

ISBN 978-0-7385-5550-8

Published by Arcadia Publishing
Charleston SC, Chicago IL, Portsmouth NH, San Francisco CA

Printed in the United States of America

Library of Congress Catalog Card Number: 2007926873

For all general information contact Arcadia Publishing at:
Telephone 843-853-2070
Fax 843-853-0044
E-mail sales@arcadiapublishing.com
For customer service and orders:
Toll-Free 1-888-313-2665

Visit us on the Internet at www.arcadiapublishing.com

To Terri and Kelly: May the road rise to meet you. May the wind be always at your back. May the sun shine warm upon your face, the rains fall soft upon the fields.

CONTENTS

ACKNOWLEDGMENTS

I am deeply indebted to many across the San Diego navy community who have added their own insights, astute perspectives, and discerning reviews to this project. The navy in San Diego is such an immense operation with such a stimulating history that no single historian has a corner on all the potential information available. To get it "right" and to seek the right amount of information and balance, one must travel many paths and seek the clairvoyance of many different experiences.

My deepest thanks go out to many in the navy's public affairs world in San Diego, including especially Capt. Matt Brown of Navy Region Southwest; and Tom LaPuzza, Alan Antzeak, and Ed Budzyna of the Space and Naval Warfare Systems Center; and to the research staff of the SeaBee Museum and the Naval Facilities Engineering Command archives at Port Hueneme.

My friends at the Maritime Museum of San Diego continuously offered accurate and perceptive advice on all things nautical in San Diego, especially Dr. Ray Ashley, Mark Montijo, Neva Sullaway, Bob Crawford, and Maggie Piatt-Walton. Likewise Katrina Pescador and Alan Renga from the incomparable San Diego Air & Space Museum and the media staff of the San Diego Aircraft Museum helped aim me in the right direction.

Staff at the U.S. National Archives, the Regional Archives at Laguna Niguel, especially director Paul Wormser and archivist Randy Thompson, and at College Park, Maryland, assisted in finding some terrific images, as did the always helpful photographic section of the Naval Historical Center at the Washington Navy Yard.

Locally I was aided by a wide spectrum of enthusiasts who contributed their time and sense of historic photography to this effort, including Bob Kyle, who contributed lavishly from the archives of the San Diego Yacht Club; fellow historian Mark Allen; Roger Clapp, force historian, Naval Special Warfare Command; and Clint Steed of the San Diego Navy Historical Association.

INTRODUCTION

Today San Diego is synonymous with American naval might: nuclear-powered carriers sit at its piers, jets flash across its skies, SEALs train on its beaches, and warships appear in nearly every picture taken of San Diego Bay. The navy is more alive in San Diego than in any other city in the nation.

Over the years, the navy's impact on San Diego has been immense and, likewise, San Diego's positive impact on the navy has been a thing of legend. The navy's presence and influence on San Diego can be felt everywhere but, importantly, always with a sense of effortless coexistence and mutual gain. The navy serves as a foundation stone for San Diego's economy and also contributes daily to the city's sense of self: its "DNA," its culture, its polish, its sophistication.

Interestingly the navy's original plans for West Coast bases did not include San Diego. In the years before World War I, as the needs for shore support for its newly modernized forces first became apparent, the navy's design for the West Coast favored its longtime presence in San Francisco Bay or a new base in Puget Sound, Washington, with little consideration for the small town of San Diego with its perilously shallow bay. The navy politely listened to San Diego lobbying but focused its attention on other alternatives, such as Honolulu and San Pedro.

Even after World War I, when a wartime glut of ships had to be distributed to a growing U.S. Pacific Fleet, the navy's grand plans for West Coast base building still placed San Diego well down on its preferred list. Only when postwar budget cuts limited new facilities was San Diego mentioned and then only as a "temporary" base for smaller ships that could navigate its shoaling bay.

But the ensuing story of happenstance, cleverness, blind luck, and changing national priorities all played to San Diego's advantage. Inexorably, year by year—through the 1920s, 1930s, World War II, and the cold war—San Diego's stock steadily rose in navy estimation. Once San Diego commanded a critical mass of naval bases, follow-on investment and further growth became preordained in cycles that today have established San Diego as the dominant American naval concentration in all of the Pacific.

Shown is a view from the deck of the nuclear-powered aircraft carrier *Carl Vinson* as she maneuvers in the turning basin off North Island in 2005 with downtown San Diego, the USS *Midway* aircraft carrier museum, and the Navy Broadway Complex in the background. (U.S. Navy.)

One

The Navy Finds San Diego

The U.S. Navy first came to San Diego's shore during the opening weeks of the Mexican War. Its mission was strategic—to seize the city and establish American rule. The draw was San Diego's incomparable bay and central location along the lengthy California coast. The aim was to provide a strategic center for supply and logistics that would support naval operations along the entire west coast of Mexico and California.

The sloop-of-war *Cyane* with its 18 guns, under the command of Comdr. Samuel Francis Du Pont, first arrived off Point Loma at dawn on July 29, 1846. Although it carried an army detachment of irregulars led by western trailblazer Maj. John C. Frémont and scout Kit Carson, the first ashore that summer day were two boatloads of sailors and marines who proceeded directly to the central plaza of the village (today's Old Town) to raise the American colors. Their welcome was warm, with many of San Diego's leading families in favor of the inevitability of U.S. rule and the attractive promise of economic gain and community stability.

As crucial as the navy was to the first American chapter of San Diego's rich history, it was not until 50 years later, at the beginning of the 20th century, that the navy was to return with any permanence. Again the attraction was both strategic and geographic in nature as the navy built both a coaling station and an experimental wireless facility at the southwestern corner of the country to support extended fleet operations. Again the city's reception was warm, and again, the promise was for economic gain and stable growth. This time, the navy engaged in the acquisition of property and negotiations for support that would set the foundation for everything that was to come.

San Diego naval history began when two boatloads of sailors and marines from sloop-of-war *Cyane* captured San Diego and raised the American flag over the town square (today's Old Town) on July 29, 1846. They were followed later in the day by irregular troops of the California Battalion led by the noted frontier explorer Maj. John C. Frémont. Warmly welcomed by most San Diegans, American forces quickly built Fort Stockton on the heights above Old Town and garrisoned the village. *Cyane* sailors would later participate in land battles near Los Angeles that secured Alta California for the United States, and the sloop helped capture and blockade ports in Baja California and along the Mexican coast. (Naval Historical Center, from watercolor by *Cyane* gunner William Myers.)

Comdr. Samuel Francis Du Pont, of the celebrated Du Pont family of American industrialists, commanded *Cyane* and ordered the first American flag to be raised over San Diego. By the time he arrived in San Diego, Du Pont had already served in the navy for 31 years and had commanded schooner *Grampus*, sloop-of-war *Warren*, and frigate *Congress*. Well-thought-of in the navy, Du Pont would later help establish the U.S. Naval Academy and would rise to the rank of rear admiral while commanding Union fleet actions against Port Royal and Charleston, South Carolina, during the Civil War. This lithograph was printed *c.* 1848 with Du Pont in the uniform of a navy commander. (Naval Historical Center.)

Frigate *Congress*, commissioned May 1842, was the last fully sail-powered frigate built by the U.S. Navy. As the Pacific Squadron flagship, she carried the flag of Commodore Robert Stockton during the Mexican War and frequently visited San Diego while the village served as the logistics base for American naval operations along the Mexican west coast. Her deep draught was ill-suited for San Diego Bay, where twice she ran aground. The first navyman to die and be buried in San Diego was Ordinary Seaman John Simpson, who fell from the fore royal yardarm of *Congress* while at anchor in the bay and was later buried at Fort Rosecrans Cemetery. *Congress* met her demise famously in 1862 in Hampton Roads, Virginia, in battle with Confederate ironclad *Virginia* the day before the remarkable engagement between *Virginia* and *Monitor*. (Naval Historical Center.)

Following the Mexican War, the navy rarely visited San Diego until ships such as the steam sloop *Hartford*, Adm. David Glasgow Farragut's famous Civil War flagship, began to visit in the years before the beginning of the 20th century. *Hartford* served as an apprentice sea-training ship along the West Coast from 1884 to 1899 and, during one visit, assigned her cutter permanently to the San Diego Naval Militia. Tradition says the *Hartford* cutter never lost a pulling-boat race (a favorite form of sport where boats from different ships would engage in spirited rowing competitions for fleet bragging rights). (Naval Historical Center.)

The army transferred land at LaPlaya to the navy in 1901 for a coaling station that the navy began constructing in 1904. This site at the mouth of the bay became the navy's first permanent base in San Diego. Due to the shallowness of the bay, only small vessels (like destroyers *Henshaw* and *Farquhar* seen above in 1925) could pull up to the pier. Others had to be replenished from coal barges while at anchor off Coronado. No sooner had Lt. Comdr. J. H. L. Holcombe assumed duties as the navy's first officer-in-charge in San Diego than the chamber of commerce began to lobby the navy to expand the base with a dry dock and repair yard. Coal can clearly be seen below in this c. 1930 photograph, laid out in long piles just east of Rosecrans Street. (Naval Historical Center.)

On May 12, 1906, San Diego's first wireless radio station, dubbed NPL for its three-letter call sign (Navy-Point Loma), rose on the crest of Point Loma, the site selected to provide maximum radio range over the ocean to the south and west of the United States. During its early years, NPL set several records for long-distance ship-to-shore radio transmission. Immediately following the San Francisco earthquake, the commander of the Pacific Squadron used NPL to relay orders to several ships to urgently proceed northward to provide aid, the first-ever use of wireless to direct fleet operations at sea. NPL would serve for over 40 years at its site on Catalina Boulevard, and in this 1924 photograph, the nearby naval coaling station can also be seen. (National Archives.)

The Chollas Heights radio transmitter site, east of downtown San Diego, was the navy's first land purchase in San Diego, bought in July 1914 for $15,000 in a land deal brokered by a helpful San Diego Chamber of Commerce. Upon its commission in 1917, the site was described as the most powerful radio station in the world and was the station that first relayed word of the Japanese attack on Pearl Harbor to Washington. The navy integrated the site (shown here in June 1958) into its worldwide radio network but demolished the towers in the mid-1990s to make room for family housing. (U.S. Navy/Naval Facilities Engineering Command.)

Two

The Great White Fleet

No single event had greater impact on solidifying San Diego's destiny with the U.S. Navy than the visit of the Great White Fleet in April 1908. This single three-day event captured the attention and imagination of the city's populace like nothing before or since, and galvanized the disorganized efforts of city leaders around a single objective: attracting the navy.

Interestingly the navy never intended to include San Diego on its world-girdling agenda. Envisioned by Pres. Theodore Roosevelt as a dramatic expression of America's global maritime power, the fleet had sailed from Hampton Roads, Virginia, in late 1907, rounded the tip of South America, and planned fleet visits only to San Pedro and San Francisco, California, and Seattle, Washington, before heading to the Orient. When it became evident that the navy's greatest battle fleet might pass within sight of Point Loma without stopping, a delegation from San Diego intercepted the fleet while it exercised in Baja California and convinced them of the value of a brief stopover.

Thousands turned out for an extravaganza of parties, receptions, speeches, parades, and banquets, and to view the imposing battleships while in port in San Diego. Ten thousand oranges were hastily arranged for immediate delivery to the ships, and city florists were put to work on the presentation and centerpiece flowers. The city's effervescent reception, gracious hospitality, and wonderful spring weather left an indelible mark on the memories of the thousands of the navy's officers and men who would rise to leadership positions in the decades to come.

Before the arrival of the Great White Fleet, the most dramatic moment in San Diego's naval history was the explosion and sinking of the gunboat *Bennington* in San Diego Bay on July 21, 1905. Officers and men of the *Bennington* gather for this photograph (above), taken in San Diego just four months prior to her sinking. Many of those pictured would be later counted among the dead and injured. Comdr. (later Rear Admiral) Lucien Young, *Bennington*'s captain, is seated at far left. Shortly after the explosion, further loss of life was averted when she was pushed from deeper water onto the city shore by tug *Santa Fe* (left). Sixty-four enlisted men and one officer perished in the explosion, and only 25 of the 200-man crew escaped unharmed. (U.S. Navy.)

A tragedy of *Bennington*'s magnitude was unheard of in a San Diego that numbered only 25,000 residents, but it helped bond San Diego to the navy in a powerful way. The dead were buried on July 23, 1905, in Fort Rosecrans Cemetery in a ceremony attended by thousands of San Diegans. Many more attended the formal dedication ceremonies on January 7, 1908, of a granite obelisk to memorialize the dead. (Naval Historical Center.)

Members of the San Diego Naval Militia are shown here *c.* 1910 in a picture taken by a local photographer. (Maritime Museum of San Diego.)

Lt. Comdr. Don Stewart leads a formation of the San Diego Naval Militia *c.* 1915 (above), and the militia was sent north to help keep the peace in San Francisco after the earthquake in 1906 (below). First authorized in 1891, the San Diego Naval Militia was a spirited addition to the community, participating in frequent local ceremonies, parades, and civic functions. The militia operated its own training ship, the fourth-rate steamer *Pinta*, as well as several training barges, and received training from warships stopping in San Diego. Stewart led the company in patrol duty along the border in 1914 to help protect the city's water supply, and the militia was mobilized into active service during World War I before ultimately being disbanded. (Maritime Museum of San Diego.)

Protected cruiser *Boston*, shown above c. 1903 at anchor off Grape Street near the current location of the San Diego County Administration Building, and protected cruiser *San Francisco,* anchored in San Diego Bay in 1891 (below), are two examples of navy ships that were assigned to the Pacific Squadron. Such ships infrequently visited San Diego before the establishment of the San Diego coaling station in 1905. Each had participated in the campaigns of the Spanish-American War but, by the beginning of the 20th century, were rapidly falling behind other more powerful warship designs. (U.S. Navy, Maritime Museum of San Diego.)

The armored cruiser *San Diego* was originally commissioned as *California* in 1907. This cruiser displaced 13,680 tons and was armed with four 8-inch and fourteen 6-inch guns, making her one of the most powerful ships of her kind in the pre-dreadnought era. *California* was renamed *San Diego* on September 1, 1914, so that the name *California* could be used for a new battleship. The *San Diego* frequently served as flagship for the commander in chief of the Pacific Fleet. While on duty with the Atlantic Fleet escorting convoys during World War I, *San Diego* struck a mine laid by the German submarine *U-156* off Fire Island, New York. *San Diego* sank in 28 minutes, but with the deaths of only six men, she was the only major warship lost by the United States during the war. An impressive Bureau of Ships model of *San Diego* and artifacts collected from the wreck can be viewed today at the Maritime Museum of San Diego. (U.S. Navy.)

This photograph shows members of the San Diego Naval Militia in 1903 or 1904, probably aboard their training ship, *Alert*. (Maritime Museum of San Diego.)

The sleek destroyer *Perry*, seen here leaving San Diego harbor *c.* 1915, is a good example of the first generation of destroyers built by the navy to combat the fast light-torpedo boat. Her open bridge, low freeboard, cramped quarters, and decks that were almost always wet all contributed to the rugged life of a crewman. (Naval Historical Center.)

Only mildly appreciated at the time, the visit of the Great White Fleet to San Diego in April 1908 is now widely regarded as the watershed event in the navy's long association with the city. The fleet was intercepted while training off the coast of Baja California by a delegation of San Diego civic officials who convinced Adm. Robert D. "Fighting Bob" Evans of the value of a stop in San Diego. Shown here are two examples of the fleet flagships, the battleship *Connecticut* (left) and the battleship *Virginia*, shown below in a striking picture. (Naval Historical Center.)

Rear Adm. Robert D. "Fighting Bob" Evans commanded the Great While Fleet during the first portion of its circumnavigation of the world and, in large part, helped place his stamp on the fleet's success. A Civil War hero who also commanded the battleship *Iowa* during the Battle of Santiago de Cuba in the Spanish-American War, Admiral Evans helped make the decision for the fleet to stop in San Diego before he was relieved of duty because of ill health. (National Archives.)

Scores of San Diegans in all manner of small watercraft flocked to view the Great White Fleet when it anchored off Coronado in April 1908. One reporter gushed, "Never before has San Diego and Coronado been the scene of such a gathering and such enthusiasm." (Maritime Museum of San Diego.)

This photograph shows the conning officer and helmsman standing watch aboard battleship *Connecticut*, flagship of the Great White Fleet. (Naval Historical Center.)

Sailors from Great White Fleet battleships parade down Broadway and past the still under-construction U.S. Grant Hotel while hundreds of spectators cheer them on. The extensive red-white-and-blue preparations that the Fleet Celebration Committee organized on little notice are clearly evident on many downtown buildings. (Maritime Museum of San Diego.)

The yachting community and the navy have always enjoyed a close social relationship in San Diego. Here millionaire Sir Thomas Lipton (center) and party, and other members of the San Diego yachting community visit Rear Adm. W. H. H. Southerland at a yachting reception in San Diego Bay in November 1912, probably aboard Southerland's flagship *California*. (San Diego Yacht Club.)

Three

The Birthplace of Naval Aviation

In 1911, the open, scrub-covered expanse of North Island represented the best place in the nation for new experiments in the fledgling science of aviation. Blessed with near-perfect winds, superior weather, protected bay-side water, open space, and a degree of security from interfering observers (as sailing or swimming were the only ways to visit the island), North Island quickly caught the attention of inventor Glenn Curtiss as a winter base for his efforts to devise the first airplane that could take off and land on water.

On January 26, 1911, Curtiss flew his untried "hydroaeroplane" from the placid waters of Spanish Bight in San Diego Bay to mark the first-ever flight of a seaplane. Immediately after this historic milestone, Curtiss intensified efforts to convince the navy to purchase his design. Ever the entrepreneur, Curtiss sweetened the offer by suggesting that he train navy pilots for free.

The navy detailed Lt. Theodore Ellyson to North Island to receive Curtiss's instruction, and Ellyson was soon designated as Naval Aviator No. 1. Shortly thereafter, naval aviation was officially born with the establishment of its first aviation squadron (consisting of two Curtiss aircraft and one of a Wright Brothers' design) at North Island.

Above, Glenn Curtiss's original hydroaeroplane rises from San Diego Bay in early 1911. Curtiss' success came from a devotion to dogged tinkering. Not dependent on any scientific study of hydrodynamics or models towed in testing tanks, Curtiss experimented with over 50 different pontoon prototypes in old-fashioned trial-and-error fashion. The first-ever flight of a seaplane (pictured below) took place in Spanish Bight on the morning of January 26, 1911, and was repeated a second time that afternoon. Once Curtiss further perfected his design with retractable wheels for land operation and trained the navy's first aviator in its operation, he was able to sell the navy two of its first three aircraft. (U.S. Navy.)

Glenn Curtiss's first marketing efforts involving the navy included offers to train naval aviators at no cost. The first aircraft bought by the navy, this Curtiss A-1 Triad (above), is shown in June 1911 on its acceptance trails. Pictured from left to right are Curtiss, Capt. Washington Chambers (in charge of naval aviation affairs for the secretary of the navy), John Towers (Naval Aviator No. 3), Theodore Ellyson (Naval Aviator No. 1), and an unidentified man. Ellyson and Towers would return two Curtiss aircraft and one Wright Brothers airplane to North Island later that year for their first season of aircraft operations. Training in Coronado continued, including the use of the Curtiss F-boat seaplane trainer shown below in Spanish Bight *c.* 1917. (Naval Historical Center, National Archives.)

One of North Island's great attributes is evident in this *c.* 1916 photograph (above)—its open spaces and flat uncluttered landing areas. This allowed any number of aircraft (such as these Army Air Corps' aircraft) to operate safely in the rather free-for-all environment of gunning engines, takeoffs, touch-and-goes, and experimental aerobatics without the encumbrance of air control direction or permissions gained from the tower. A seaplane trainer is shown below, also *c.* 1916, readying for engine start in Spanish Bight. (National Archives.)

With the arrival of the navy's first aircraft carrier, *Langley* (seen here inbound in the San Diego channel and alongside her berth at North Island), in San Diego on November 29, 1924, a new era for the city began. Converted from a large collier by the erection of an awkward flight deck and other features to support both carrier aircraft and seaplanes, *Langley* would routinely operate within sight of either Coronado Beach or Point Loma while pilots trained in carrier procedures and developed carrier-operating doctrine that, in some cases, are still in use today. In 1925, the first squadron trained to operate from a carrier completed their deck qualifications, and that same year, *Langley* conducted the navy's first night carrier landing. (San Diego Air & Space Museum, National Archives.)

Naval aviation activity in San Diego accelerated rapidly after World War I as squadrons of Battle Fleet aircraft concentrated their training on North Island. A Curtiss R-6L floatplane experiments with an early air-launched torpedo while training in San Diego Bay in September 1920. (Naval Historical Center.)

Adm. Hilary P. Jones shakes the hand of Lt. John D. Alvis next to North Island's large seaplane hangar as Capt. Albert Marshall looks on in April 1923. Alvis is nattily attired in the aviation dress uniform of the day. (Naval Historical Center.)

North Island was the scene of many of the navy's experiments with lighter-than-air craft. Three different types of airships (above) are shown next to North Island's newly completed dirigible hangar in February 1921. Of note, the air station's only observation tower for field operations is perched at the hangar's peak. In October 1924, the massive *Shenandoah* (below), the first zeppelin-size airship built in America and the first rigid airship ever seen on the West Coast reached North Island's mooring mast after a transcontinental flight from Lakehurst, New Jersey. (Naval Historical Center, National Archives.)

The navy's first phase of construction at North Island included the tower of the administration building and two seaplane hangars at water's edge of Spanish Bight, shown here in June 1923. Much to the navy's credit, North Island architecture was heavily influenced by the attractive Spanish Colonial Revival style used by Bertram Grosvenor Goodhue for the principal exposition buildings in Balboa Park. This architecture enjoyed tremendous popularity during the period and helped the navy blend in with regional styles. (U.S. Navy/Naval Facilities Engineering Command.)

Curtiss N-9 seaplane trainers line up early on a clear morning in 1918. (San Diego Air & Space Museum.)

The army's Rockwell Field stands at the top of the photograph of North Island in 1927. Carrier *Langley* and navy administration buildings are at the bottom, with navy airplane hangars at the middle right. A wooden landing deck for practice carrier landings can be seen in the center of picture. (U.S. Navy.)

It was an active day in the mid-1930s with a *Northampton*-class heavy cruiser moored mid-stream while a crew prepared a Douglas PD-1 flying boat for takeoff at North Island. (National Archives.)

Langley demonstrates its ability to launch a Douglas DT-2 aircraft while moored at North Island in 1925. (National Archives.)

Rear Adm. Joseph Reeves (fourth from right, front row), commander of aircraft for the battle fleet, is shown with his staff on the steps of the North Island administration building in the fall of 1927. To most San Diegans he was "Bull" Reeves, and through his leadership, dramatic strides were made in naval aircraft tactics in operations and experiments in and around San Diego. (National Archives.)

Six new P2Y-1 Ranger patrol seaplanes, the navy's first long-range patrol aircraft, were flown to San Diego in 1933 to train for the first-ever nonstop squadron flight from the mainland to Hawaii. These aircraft from VP-10 under the command of Lt. Comdr. Soc McGinnis accomplished this feat in 24 hours and 35 minutes on January 10, 1934. (National Archives.)

Four

San Diego Finds the Navy

Historical coincidence, perfect timing, and event alignment are frequently the stuff of fictional novels where authors manage plot, pace, and personalities. The appearance of similar precision in the realm of non-fiction history is rare but, when found, should be admired and valued. Such an occurrence united San Diego and its navy across an exciting decade beginning in 1912.

At the time, San Diego was at a crossroads, caught in a stirring debate between those stressing the imperatives of economic and industrial growth, and others defending the region's natural attributes. Most community leaders, dazzled by dreams of prosperity and growth, supported agendas of industrialization and expansion. A smaller, but vocal faction called for a different picture of San Diego, one of pristine orange and avocado groves, warm beaches, and healthy sunshine that would spur tourism and "soft" growth.

Guided by San Diego congressman William Kettner and others, the navy soon was seen as the perfect compromise for San Diego. If the navy could be lured to San Diego, it would funnel money into the city to spur growth without the need for industrialization; it would help develop the city's primary natural attribute—its bay—and would help bring to San Diego a progressive, paycheck-in-the-pocket population. It was close to the perfect solution, and, beginning in 1912, San Diego set its strategy in motion.

With the end of World War I, the navy redistributed its fleet by sending a large number of ships to reconstitute the Pacific Fleet. Although navy planners preferred San Francisco Bay as the primary West Coast base, facilities there were not ready to receive all the ships the navy intended to send westward, and shallow draught ships like destroyers were assigned temporarily to San Diego and larger ships to San Pedro. This sudden glut of ships quickly overwhelmed San Diego's berthing facilities, causing the navy to moor them either in mid-stream (as was done with Destroyer Division 39, shown above in a 1921 photograph) or at any convenient location along a bay shoreline that was undergoing reclamation (such as near the Ferry Landing *c.* 1919, below). (Naval Historical Center, National Archives.)

Local congressman William Kettner helped find investors during World War I to open a shipyard on leased land south of downtown to build two ships, *Cuyamaca* (above) and *San Pascual*, experimentally made of concrete for the war effort. Shown below are steel bar strenghtheners on the interior of the ship. To protect the government's investment in construction machinery once the ships were completed, Kettner then helped broker a deal with the San Diego City Council in which the shipyard land was donated to the navy in September 1919. Kettner also sponsored an appropriation in Congress for site improvements for the "repair of destroyers and other craft." These deft maneuvers formed the basis for what would become the San Diego Destroyer Base at 32nd Street. (Maritime Museum of San Diego.)

Destroyer *Fairfax* (DD-93), shown above off the California coast in May 1918, was typical of the World War I-era 4-piper destroyer that was homeported en masse in San Diego following the war. Mostly placed in reserve and moored in tightly packed conditions at the 32nd Street Naval Base—nicknamed "Red Lead Row"—many of these ships (below) were returned to duty as conditions merited and resulted in a new concentration of ship-repair trades in San Diego that helped spark a rebirth of the ship-repair economy. Beginning in 1938, many of these 4-stack destroyers were recommissioned to counter U-boats or were loaned to Allied navies. (Naval Historical Center.)

This photograph shows Adm. William Banks Caperton, commander of the Pacific Fleet, and his staff officers aboard his flagship, armored cruiser *San Diego*, in San Diego Bay, January 1917. (Naval Historical Center.)

By the 1920s and early 1930s, San Diego's destroyer roots had been well established. The destroyer base slowly evolved into a major repair and maintenance location with destroyer tenders and ranks of destroyers in reserve. Then-Capt. Chester Nimitz commanded the base in 1931–1932 and quartered his family aboard tender *Rigel* (shown here in the middle of the 32nd Street waterfront), where the ship's chart house was converted to his son's bedroom. (Maritime Museum of San Diego.)

Destroyers *Cassin* and probably *Cummings* lay a smokescreen while at full speed *c*. 1919 and demonstrate the verve and dash that were associated with duty aboard destroyers of their day. (National Archives.)

This photograph shows officers and men (and a pet dog) of destroyer *Henshaw* and other ships of Destroyer Division 23 moored in San Diego during June 1920. (Maritime Museum of San Diego.)

This photograph shows a nest of destroyers around tender *Rigel* in 1929 or 1930. Active destroyers *McCawley*, *Zeilin*, and *John Francis Burns* are being retired with their crews and equipment shifted to reserve destroyers *Perry*, *Philip*, and an unknown destroyer tied up immediately on the tender's port side. Thirty-four destroyers were exchanged in this manner, providing an economic stimulus for San Diego but significantly disrupting the lives of fleet crewmen. (Maritime Museum of San Diego.)

With naval operations rapidly expanding in San Diego by early 1920, the navy recognized a need to bring a flag officer to the city to coordinate its widespread activities. Rear Adm. Roger Welles, a battleship officer, was assigned as the first commander of the new 11th Naval District in San Diego. (Naval Historical Center.)

U.S. congressman William Kettner (right, in Panama hat) and navy secretary Josephus Daniels (center) are shown here aboard the battleship *New Mexico* in San Diego Bay during September 1919. The pair teamed in a manner that consistently favored San Diego's naval expansion in the years around World War I. Kettner, a member of the San Diego Chamber of Commerce, adroitly aligned Congressional appropriations and naval support for the expansion of San Diego naval facilities until a critical mass of bases was established. (Naval Historical Center.)

This photograph shows the battleship *Texas* anchored off North Island in 1929 with the San Diego waterfront and the new El Cortez Hotel in the background. Contrast this picture of the waterfront with the photograph on page 8. (Maritime Museum of San Diego.)

Three light cruisers of the *Omaha*-class swing at anchor on a tranquil morning in San Diego in December 1934. (National Archives.)

Facing the need to rapidly expand its recruit training to meet the requirements of World War I, the navy accepted San Diego's proposal to lease the vacant buildings of the Panama-California Exposition in Balboa Park for a dollar a year in May 1917. Navy recruits (above and below) stand in ranks in Plaza de Panama in 1917, during the time when it served as the site for the first San Diego Naval Training Station. (National Archives.)

Nearly 4,000 sailors passed through this temporary boot camp in Balboa Park to receive a wide variety of instruction, such as in this January 1918 boxing class. With so many sailors at the training station, the navy also established a modest dispensary near the site, a move that led to the establishment of the Balboa Naval Hospital. (U.S. Navy.)

During October 1918, with help from the San Diego Chamber of Commerce, the navy inspected different sites for a permanent new training station. Rejecting locations on the Silver Strand and near the current location of Sea World on Mission Bay, the navy finally settled on this 135-acre site (partially underwater) at Loma Portal, which was provided at no cost after the city council raised $280,000 by public subscription donations for the land. (U.S. Navy/Naval Facilities Engineering Command.)

With great fanfare, the naval training station was commissioned on Navy Day, October 27, 1923, with William Kettner, San Diego civic leaders, and Capt. David Sellers (the station's first commanding officer) sharing the spotlight (above). Recruit training started almost immediately, and by the time the aerial photograph shown below was taken in September 1924, more than 350 sailors were aboard for advanced training at the Service School Command, including radiomen, yeomen, buglers, and musicians. The command also was home to the navy's only school for electricians. (Naval Historical Center, U.S. Navy/Naval Facilities Engineering Command.)

Recruits in the late 1920s at the naval training station learn that there is a proper way, as well as an improper way, to stow their belongings in their sea bags and hammocks. Aboard ship, hammocks were stowed in the overhead during the day and provided a ready stowage area for any amount of personal gear. (National Archives.)

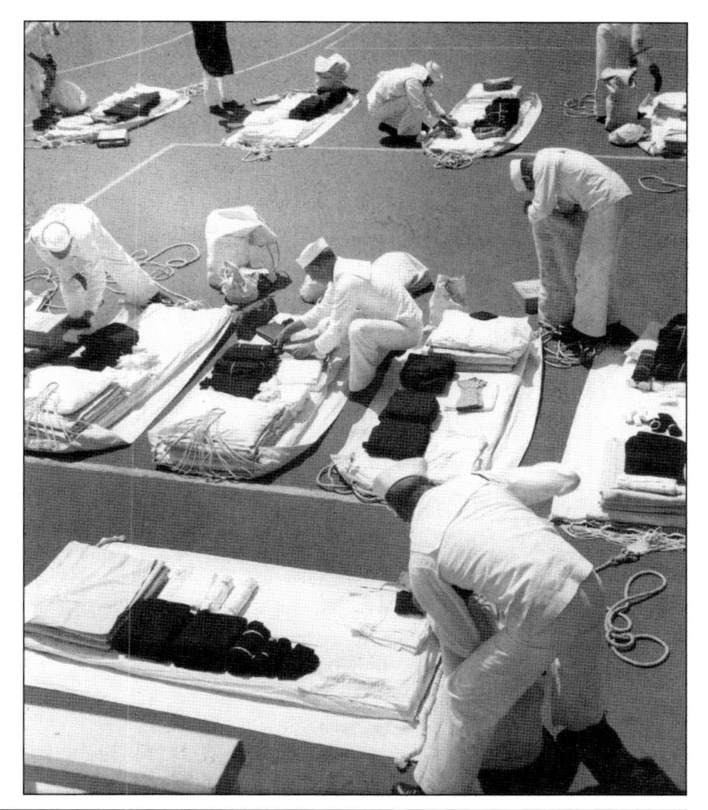

Navy recruits, fresh off a bus or train, line up at the main gate of the naval training station in 1923 to be inducted. (National Archives.)

This photograph shows a traditional review of graduating recruit companies at the naval training station *c.* 1935. Although several naval training sites were closed during the 1930s for economic reasons, the San Diego Naval Training Station witnessed a period of expansion because of its excellent training weather and proximity to the Pacific Fleet. Consistent infusion of New Deal recovery funds allowed for new buildings, streets, power stations, lawns, and tennis courts. By 1939, facilities that had been improved stood ready to receive a sudden flood of new recruits needed to fill the rapidly expanding fleet. By the end of the decade, the training station could accommodate 5,000 recruits and 1,000 additional ratings in training. (Naval Historical Center.)

Not only did the San Diego Naval Training Station train new navy recruits in the well-known boot camp environment, it also hosted a series of specialty schools that taught apprentice and journeymen courses in specialized ratings. Here, *c.* 1934, navy yeoman students are instructed using the traditional heavy typewriter of the era. (Naval Historical Center.)

The old Balboa Naval Hospital was also a mainstay of naval activity in San Diego. This interior detail photograph of the hospital shows why it was so popular with both the navy and the citizens of San Diego. (U.S. Navy/Naval Facilities Engineering Command.)

Postcards such as these were sold at stores on Broadway and outside the gate of the San Diego Naval Training Station. They were favorites among the sailors in the 1920s and 1930s to help maintain their correspondence home. (San Diego Navy Historical Association.)

The naval hospital, located on Balboa Park's Inspiration Point, was the last of the four major facilities the navy would build in San Diego during the Kettner era. An outgrowth of a small dispensary that supported the temporary World War I naval training station, critics decried the buildings for being too austere when they opened in 1922. This led the navy to endorse an aggressive program of landscaping that helped the hospital fit with the other components of Balboa Park and ultimately led to its nomination to the U.S. Register of Historic Places. (National Archives, U.S. Navy/Naval Facilities Engineering Command.)

This photograph shows the aircraft carrier *Saratoga* at anchor in the bay *c.* 1933 with carrier *Langley* and seaplane tender *Wright* at the North Island pier. A causeway extended from Coronado's Fourth Street past the city's old golf course and across Spanish Bight to North Island, where the original guardhouse can still be seen today. (National Archives.)

Squadron aircraft at North Island line up for Saturday inspection in May 1932. Many of these are assigned to the carrier *Saratoga*. (National Archives.)

A formation of Martin T3M-2 torpedo planes assigned to VT-2B aboard carrier *Saratoga* lumber over Coronado Heights and South Bay c. 1928. The T3M-2 was the first torpedo-carrying aircraft to be developed specifically for carrier operations. Tactical exercises and experimentation off San Diego with these aircraft significantly enhanced the effectiveness of torpedo attack, the primary anti-ship offensive capability of naval aviation. (San Diego Air & Space Museum.)

Squadron members stand in front of a Curtiss R-6L seaplane at the North Island seaplane hangars during July 1920. The R-6L was used for a variety of purposes, including experimentation with the first air-launched torpedoes, and was the first American-built aircraft used overseas during World War I when it was assigned to patrol duties in the Azores. (Naval Historical Center.)

A Boeing F4B-2 fighter assigned to fighter squadron VF-5B aboard aircraft carrier *Lexington* flies over Ocean Beach and Mission Bay in 1932. (San Diego Air & Space Museum.)

The odd-looking but ultimately successful Loening OL-8 observation aircraft assigned to utility squadron VJ-1F at North Island is wrestled up the ramp *c.* 1928. OL-8s had three seats, the third frequently used by an aerial photographer. This aircraft is flying a two-star flag, which indicates the presence of a rear admiral on board, possibly the commander of aircraft for the battle fleet, who was based in San Diego. (San Diego Air & Space Museum.)

This 1931 photograph of North Island's rapidly expanding infrastructure includes the base's first seaplane hangars to the left and a fully loaded carrier *Langley* in the foreground. In an era before paved runways, echelon arrows painted on the ground directed pilots who were taking off or landing toward the wide-open parts of the field and away from parked aircraft on the tarmac. (San Diego Air & Space Museum.)

The army purchased the 2,100-acre Miramar Ranch on Kearny Mesa during World War I and developed an infantry-training center named Camp Kearny. Later the Camp Kearny parade field was used as a navy dirigible base, and by 1936, the navy had established an unpaved emergency landing field there. Here three F2F-1 fighters of fighter squadron VF-2 from carrier *Lexington* fly over an undeveloped portion of Miramar Ranch in July 1939. (Naval Historical Center.)

Destroyers pack Broadway Pier during a weekend in October 1940, offering liberty-bound sailors a direct shot to Broadway. Note Lane Field (baseball) to the far right, the Naval Supply Center, and the North Island Ferry landing, operated by the Star and Crescent Boat Company, in the left foreground. Broadway Pier was known for its ornate facade and picturesque palm stands in the middle of Harbor Drive. (Naval Historical Center.)

On February 8, 1934, the giant navy airship *Macon*, shown above and below, flew over downtown San Diego "and stopped all traffic across town," according to the *San Diego Union*. As big as a battleship, she operated from a temporary mooring mast at the Camp Kearny airship facility (today's Marine Corps Air Station, Miramar) in 1934 while engaged with *Saratoga* squadrons to develop long-range scouting tactics. Tragically, the *Macon*, the last of the navy's giant dirigibles, was lost in an accident off California's Point Sur in February 1935. (National Archives, Navy Historical Center.)

In 1911, two U.S. Navy submarines, *Grampus* and *Pike*, introduced San Diego to the new world of the submersible warship and, in the process, frightened ferry boat captains and crew during practice dives in San Diego Bay. Until the mid-1930s, submarines were based in San Francisco Bay and at San Pedro but, in an efficiency measure, were ultimately concentrated at San Diego's 32nd Street Naval Station. (Naval Historical Center.)

In the years prior to World War II, San Diego was teaming with Pacific Fleet submarines. Here a nest of newly commissioned subs is tied to a tender moored in San Diego Bay *c.* 1940. A *Yorktown*-class aircraft carrier is at the carrier pier on North Island. (Naval Historical Center.)

Naval history was celebrated in January 1933 with the visit of frigate *Constitution*, known as "Old Ironsides," to downtown San Diego as part of a cruise along the Atlantic and Pacific coasts to commemorate her restoration, funded through donations from schoolchildren and patriotic groups. Crowds of San Diegans waited hours in long lines to go aboard the naval icon. (Maritime Museum of San Diego.)

Officers and men of the light cruiser *Omaha* pose for this formal picture, taken in San Diego Bay *c.* 1936. (Maritime Museum of San Diego.)

It is hard to find a picture of San Diego Bay that does not show evidence of the positive impact the navy has had on the community, including this 1936 aerial photograph that shows North Island development, the restored shoreline behind the Embarcadero quay wall that one day would hold the San Diego Convention Center, two navy ships at anchor in the harbor, and the Coronado ferry plying the bay. Note the *Star of India*, San Diego's most famous maritime symbol, moored at the quay—originally towed to San Diego by the navy minesweeper *Tern*. (City of San Diego.)

One of the original motivations to lure the navy to San Diego was to seek federal support for harbor dredging in order to attract the commerce of deep-draught merchant ships. By the 1930s, several different phases of harbor improvement had already been accomplished, not only deepening the bay for large navy aircraft carriers and battleships, but also helping spur the San Diego economy. Dredging also provided fill material to help shape and expand the shoreline, clearly evident here in this 1940 photograph, especially around North Island, Coronado, Shelter Island, Spanish Landing near the airport, and downtown near Broadway Pier. (U.S. Navy/Naval Facilities Engineering Command.)

Five

World War II

World War II changed San Diego forever. Upon the foundation that previous naval investment had established in the 1920s and 1930s rose a sweeping defense establishment that would dominate nearly every part of San Diego city life for four long years. San Diego naval bases expanded dramatically, and new airfields, training sites, and laboratories proliferated across the countryside. Industrial production skyrocketed, most notably in ship repair, fleet support services, and aircraft production. Defense workers and navy sailors streamed into the city from every state, straining the city's utilities, housing, and infrastructure. It was jarring, uncomfortable, and frightening, while, at the same time, it was exciting, captivating, and inspirational.

Everywhere the tempo was high. The navy controlled all harbor operations, local universities focused on war technologies, rationing impacted kitchen pantries, and entertainment venues operated around the clock. Piston engines droned overhead while trolleys and buses clanged along downtown streets largely devoid of automobiles. The San Diego Naval Training Station and Balboa Naval Hospital both doubled their operations in the first year of the war and then doubled them again in the second.

Many San Diegans today remember the vibrancy of the wartime city with barely disguised fondness—the chance for new acquaintances, the patriotic verve of national service, the excitement of headlines, and the candid realization that, if not for the exigencies of war, they might never have discovered San Diego.

The incomparable PBY Catalina flying boat became synonymous with naval aviation in San Diego, where Consolidated Aircraft built over 2,300 between 1936 and 1945, elevating San Diego to the top tier in the nation's aerospace industry. The distinctive and versatile PBY was noted for its long-range search capabilities but was also used for the search-and-rescue of downed pilots, nighttime bombing, and supply and transport. Here a PBY-5A flies over Torrey Pines in March 1942 (below), and Catalinas pack the North Island flightline in 1938 (above), where aircraft usually were delivered from the Consolidated Aircraft plant at today's Lindbergh Field. (San Diego Air & Space Museum.)

Two auxiliary runways were installed at Camp Kearny in 1940 and 1941, and when these were completed in February 1943, the navy formally established the Naval Auxiliary Air Station (NAAS), Camp Kearny (shown below). A companion airbase located north of Camp Kearny helped support marine units and was designated the Marine Corps Air Depot, Miramar in September 1943. By the end of the war, NAAS Camp Kearny was supporting squadrons of navy fighters and PB4Y-2 Privateer patrol bombers such as "The Kamikaze Miss," shown above in 1945. (Above, San Diego Air & Space Museum; below, U.S. Navy/Naval Facilities Engineering Command.)

Wartime construction exploded across San Diego County. Nearly overnight, bases were constructed for training purposes and to house the hundreds of thousands of military personnel streaming into San Diego. The navy purchased 130,000 acres north of San Diego and began construction in March 1942 on what would become the nation's largest U.S. Marine Corps base, Camp Pendleton. This June 1944 photograph (above) shows construction and reclamation work around the Del Mar Boat Basin that would service a host of amphibious craft. The navy also constructed a series of nearby airfields to augment air training at North Island. Fields were constructed at Del Mar, Sweetwater Lake, Salton Sea, Ramona, Otay Mesa, and this site (below) in the desert at Holtville. (Above, U.S. Navy/Naval Facilities Engineering Command; below, National Archive.)

The army originally leased the site of Ream Field, located between Imperial Beach and the Mexican border, in 1918, but the navy began using the field in the 1920s for weapons training and carrier-landing practice. By the time this photograph was taken in August 1958, the field was well on its way to becoming the principal navy helicopter base in the Pacific. (U.S. Navy/Naval Facilities Engineering Command.)

Due to sailing restrictions in the bay during the war, the San Diego Yacht Club pioneered the use of Penguin-class sailboats for competitive racing. Kettenburg Boat Works was a prime builder, and after the Naval Training Center bought 30 Penguins in 1942–1943, frequent contests were held between local and navy yachtsmen to boost morale and foster recreation. (San Diego Yacht Club.)

The waterfront around Broadway Pier swirls with activity and teams with marines, sailors, and their loved ones in December 1943 as troops of the 4th Marine Division wait in formation to board. These ships (seen in both images) would form the primary assault group for Operation Flintlock, the amphibious attack on the Japanese-held Marshall Islands and Kwajalein Atoll that occurred January 31–February 3, 1944. Among the ships that can be identified are *Aquarius* (AKA-16) and *DuPage* (APA-41). (National Archives.)

The navy first came to Coronado Heights, north of Imperial Beach, in 1920 to establish a radio navigational beacon site. Just after Pearl Harbor, the army began construction on two large concrete gun emplacements for 16-inch harbor defense guns at what they called Fort Emory. At the same time, navy activities were devoted to the development of highly secret radio-direction finding and radio intelligence facilities. The navy radio intelligence intercept cell at Imperial Beach was called Station ITEM (for the phonetic letter I) and was part of a Pacific-wide network of stations that could triangulate enemy ship locations. By war's end, Coronado Heights boasted a large communications and cryptology training facility and, for years, featured the eye-catching Wullenweber circular antenna array. (Above, National Archives; at right, U.S. Navy.)

The San Diego Naval Training Station began a major expansion project in 1939 and, by the beginning of World War II, was increasing its throughput at a prodigious rate. By September 1942, the station had already reached its wartime peak of 40,000 sailors, including 25,000 recruits engaged in training that routinely spanned nine hours a day, seven days a week. (Maritime Museum of San Diego.)

The gunnery school, auditorium, and firefighters school at the Naval Training Center are shown still in their wartime paint scheme in April 1945. (U.S. Navy/Naval Facilities Engineering Command.)

The dramatic wartime enlargement of the Naval Training Center can be seen in this May 1949 aerial photograph, which includes the eye-catching expansion of Lindbergh Field on fill land that still preserved the boat channel between the airport and the naval center. (U.S. Navy/Naval Facilities Engineering Command.)

Women Accepted for Volunteer Emergency Service, the official title of women reservists in the navy, was better known by its acronym, WAVES. The first groups of WAVES in San Diego arrived at the naval training station for indoctrination in May 1943 and are reviewed here by secretary of the navy Frank Knox a month later. (Naval Historical Center.)

The first quarters for WAVES in San Diego were built in Coronado during 1942 on land near the Ferry Landing between E and G Streets. This bay-front property is now some of the priciest property in all of San Diego—quite a deal for some of the navy's first women members. (National Archives.)

Wartime leaders made a point of traveling to San Diego to encourage sailors and marines, and to work with San Diego civic leaders to mitigate the negative impact of the massive influx of navy sailors and defense workers that stressed nearly every facet of city life. Here Pres. Franklin Roosevelt visits the naval hospital in 1942. (Naval Historical Center.)

Navy nurses undergoing training in San Diego at the naval hospital march through Balboa Park c. 1944. (National Archives.)

The navy established its own policing force, the Shore Patrol, in San Diego during 1942 to assist in controlling unruly sailors on leave and to create a more positive atmosphere of navy-civic cooperation. San Diego entertainment venues during the war ran the spectrum from round-the-clock dance halls, to tattoo parlors, penny arcades, and peep shows on Broadway. One never wanted to run afoul of the stern hand of law enforcement personified by no-nonsense chief Harry Morris. (Maritime Museum of San Diego.)

With the navy's total control of the port and with nearly all private shipping devoted to the war effort, the San Diego private shipbuilding industry sought new opportunities to contribute to meet navy requirements. Kettenberg Boat Works won a lucrative navy contract to build 34-foot plywood plane rearmament boats (above) to carry bombs and torpedoes to seaplanes. Likewise, when local trawlers and tuna clippers were taken up into naval service as auxiliary antisubmarine and patrol craft, Kettenberg designed a new class of small replacement fishing boats (characterized by *Baby Doll*, below) for the fishing industry. (Maritime Museum of San Diego.)

IMPERIVM NEPTVNI REGIS

TO ALL SAILORS WHEREVER YE MAY BE: and to all Mermaids, Whales, Sea Serpents, Porpoises, Sharks, Dolphins, Eels, Skates, Suckers, Crabs, Lobsters and all other Living Things of the Sea

GREETING: Know ye, That on this 21st day of Aug. 1942 in Latitude 0000 and Longitude 175°-00'W. there appeared within Our Royal Domain the U.S.S. San Diego bound South for the Equator and for wherever the enemy may be

BE IT REMEMBERED

That the said Vessel and Officers and Crew thereof have been inspected and passed on by Ourself and Our Royal Staff And Be It Known: By all ye Sailors, Marines, Land Lubbers and others who may be honored by his presence that

GEORGE H. HORTON G. H. H. Cox. USN

having been found worthy to be numbered as one of our Trusty Shellbacks he has been duly initiated into the

SOLEMN MYSTERIES OF THE ANCIENT ORDER OF THE DEEP

Be It Further Understood: That by virtue of the power invested in me I do hereby command all my subjects to show due honor and respect to him wherever he may be.

Disobey this order under penalty of Our Royal Displeasure

Given under our hand and seal this August 21, 1942.

Davey Jones
His Majesty's Scribe

Neptunus Rex
Ruler of the Raging Main

By His Servant ______ USN

SIGILLVM OFFICII NEPTVNI REGIS

Signed by Davey Jones and Neptunus Rex, this Ancient Order of the Deep signifies that one George H. Horton of the cruiser *San Diego* appeared within the "Royal Domain" during a crossing-the-line ceremony on August 21, 1942, at longitude 175-00W, latitude 00-00 bound south for the equator and for "wherever the enemy may be." (Maritime Museum of San Diego.)

The World War II war effort dramatically expanded San Diego's naval infrastructure and the county's population. To solve what he called an "impending emergency" of water shortages, Pres. Franklin Roosevelt cut through red tape and directed the navy to construct a new aqueduct into the county. Completed in December 1947, this aqueduct guaranteed an increased water supply to San Diego and helped spur economic prosperity well into the 1950s and 1960s. (U.S. Navy.)

The light cruiser *San Diego* was one of the most decorated ships of World War II, earning 15 battle stars for service in battles from Guadalcanal to the final offensive against the Japanese home islands—all without losing a single crewman to hostile action. She held the distinction as the first major Allied ship to enter Tokyo Bay and led the occupation of the Yokosuka Naval Base (pictured here). She visited her namesake city twice, once shortly after her shakedown training in 1942 while escorting carrier *Saratoga* to the battlefront and for the final time when she entered the harbor to a enthusiastic citywide welcome in 1945 two months after the surrender. (Naval Historical Center.)

San Diego welcomes the news of Japan's surrender on August 14, 1945. Thousands join the downtown freewheeling celebrations remembered by many for decades, while those aboard destroyer *Kendrick* at the naval repair base gather for a commemorative photograph. (National Archives.)

Light cruiser *San Diego* sailors enjoy an evening of liberty during the ship's only break from the war at Mare Island Naval Shipyard in May 1944. (Maritime Museum of San Diego.)

A simple sign indicates the enthusiastic citywide support for all sailors. The Ballast Point greeting was in place in December 1945 to welcome the tens of thousands of servicemen who were flowing back through San Diego on their way home from the war. (National Archives.)

Six

Technology Capital of the Navy

San Diego's current standing in the world of research and development is impressive. The city is the technology capital of the navy: the hub of a vast array of intricate engineering laboratories and headquarters for much of the navy's technical community.

Developments beginning in the 1920s helped spur much of the concentration of technical and engineering talent in San Diego that would help form the foundation for future growth. Many of the navy's aeronautical advances in aircraft, weapons, material, training, and tactics were initially realized in San Diego during the 1920s and 1930s. The 1930s also saw important breakthroughs in the use of underwater sound to detect submarines, and nearly all of the navy's early work with sonar in antisubmarine operations had its roots in San Diego.

In the first years of World War II, naval partnerships with the University of California and Scripps Institute of Oceanography spawned a host of wartime laboratories along Point Loma specializing in wave and beach forecasting techniques for amphibious landings, underwater sensors, radar, advanced communications, improved weaponry, and signals intelligence. This focal point of navy laboratories matured and broadened during the years of the cold war, aided by a dimension that few San Diegans ever saw—the navy's extensive and unmatched concentration of offshore test ranges and operating areas. Today San Diego's technological contribution to American defense needs includes groundbreaking work with unmanned aircraft, robotics, data networks, intelligence sensors, and underwater sound.

On June 1, 1940, the Navy Radio and Sound Laboratory (NRSL) was established. It was the navy's first laboratory on the West Coast and was the vanguard of the navy's technology concentration on Point Loma. This 1942 photograph shows some of the NRSL's first buildings grouped near the NPL radio station (see page 15). Some were painted in a camouflage paint scheme common in the first years of the war. (U.S. Navy.)

The military reservation on Point Loma (shown here in 1939) had been in place since the mid-1800s, and its relatively undeveloped space provided ready land on which the navy rapidly developed its laboratory complexes at the beginning of World War II. (U.S. Navy/Naval Facilities Engineering Command.)

In 1939, the navy's first sound school was originally established at the naval station. In 1943, it expanded to this wedge-shaped spot of land next to the Naval Training Center and Harbor Drive, where ships of the West Coast Sound Training Squadron could berth at adjacent piers and augment classroom training in the new technology of sonar. San Diego has since been the navy's center for antisubmarine training and tactics development. (U.S. Navy/Naval Facilities Engineering Command.)

During World War II, the Navy Radio and Sound Laboratory in San Diego was a center of research on underwater sound transducers and acoustic homing torpedoes, and much of its fieldwork was conducted in Sweetwater Reservoir, which was deeper than San Diego Bay and free of background noise. (U.S. Navy/SPAWAR Systems Center [SSC] San Diego.)

A fixture on the Point Loma skyline since 1945, the antenna model range provides the means to check the dynamics and science of shipboard antennas and sensors. Model ships sit in a pond of water while antenna and transmitter elements are arrayed on a gantry arm above. The models are commonly made of brass, and old models of vintage warships have made interesting collector items at locations around town. (U.S. Navy/SSC San Diego.)

San Diego naval laboratories have a long association with oceanography and marine-engineering studies, beginning with their close relationship with the Scripps Institute of Oceanography during World War II. The deep-submersible *Trieste* was used for a broad range of underwater experiments in deep water, as well as for tests of sonar instruments. In January 1960, *Trieste* descended into the Pacific Ocean's Challenger Deep to a record 35,800 feet. (U.S. Navy/SSC San Diego.)

Surprisingly, beginning in the 1950s, San Diego was the center for the navy's arctic research efforts that centered on under-ice sonar, hull strengthening, navigation, and weapons. When *Skate* surfaced at the North Pole in March 1959, she used sonar pioneered in San Diego that permitted the nuclear-powered submarine to penetrate the ice pack and to surface through several feet of ice. (U.S. Navy/SSC San Diego.)

San Clemente Island and the naval operating areas that surround it provide the navy with technology resources unmatched anywhere else in the world. The navy established an instrumented underwater test range off the island in 1951 to test new torpedoes, such as this Mk46 lightweight torpedo (above), and to conduct antisubmarine exercises. Numerous weapons systems have also been tested in the relative secrecy of offshore waters, such as the land-attack Tomahawk missile and the earlier 1,200-mile Regulus II cruise missile, shown here in a test launch from the guided-missile test ship *King County* c. 1959. (Above, U.S. Navy/SSC San Diego; below, National Archives.)

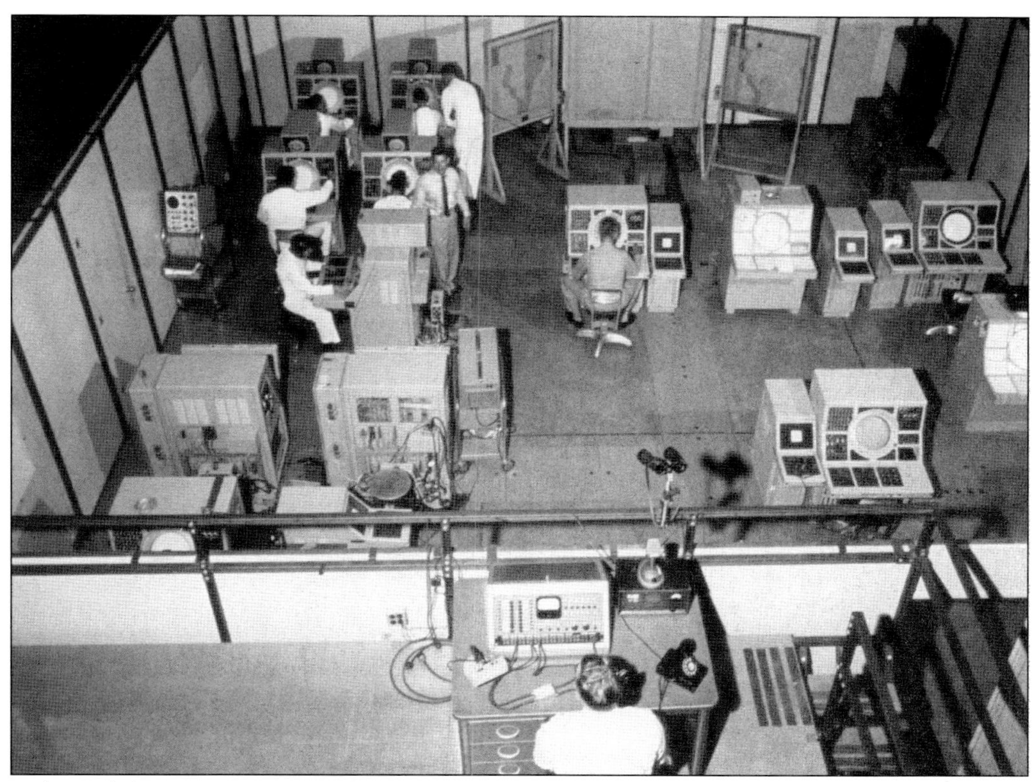

Delivered for testing in 1958, the Navy Tactical Data System (NTDS), largely overseen and developed in San Diego naval laboratories, consisted of computers, stored programs, specialized displays, and digital data links. This breakthrough use of technology revolutionized naval tactical operations and helped shift the navy from analog to digital data processing. (U.S. Navy/SSC San Diego.)

Research on tactical data systems in San Diego also includes new display technologies, such as this Tactical Flag Command Center (TFCC) mockup for aircraft carriers and command ships, as well as data security, network operations, and connectivity between ships, aircraft, and unmanned aircraft—all technologies that are in a constant state of change. (U.S. Navy/SSC San Diego.)

Undersea surveillance work at San Diego's Naval Undersea Center led to the development of a mobile towed array called the Surveillance Towed Array Sensor System (SURTASS). *Stalwart* was designed specifically to deploy SURTASS to provide extremely long-range detection of Soviet nuclear submarines. (U.S. Navy/SSC San Diego.)

On April 4, 1960, Polaris, the navy's first ballistic missile system, was successfully fired underwater for the first time from a test launcher off San Clemente Island. This program led to the Poseidon and Trident missiles that were fired from nuclear-powered submarines, such as the *George Washington*, that aided the nation's strategic deterrence. (U.S. Navy/SSC San Diego.)

Seven

The Cold War

The U.S. Navy developed its presence in San Diego in a big way in the 1920s and 1930s. The demands of the war effort of World War II had seen a fundamental transformation of the city with huge influxes of population and industry, and a navy that, in many ways, imposed itself with little thought of its impact. Worldwide victory in 1945 brought broad demobilization and the first glimmer of another new phenomenon—as the navy began to close hundreds of bases across the country, it concentrated the remaining parts of its establishment that it wanted to keep in a small number of critical locations, and San Diego was one of those. As other bases closed, San Diego benefited.

The nation's broad strategy for the cold war was to "contain" the Soviet superpower, and the navy's contribution to that strategy moved naval task forces to the forward flanks of the Soviet Union. The fleet reshaped itself for a protracted, drawn-out overseas presence.

This change in strategy benefited the San Diego–U.S. Navy relationship in ways not immediately obvious. To maintain a consistent naval force level overseas, the fleet invented a cyclical deployment pattern, nominally six months in duration, and quickly aligned all manpower policies, training schedules, and support services to meet this imperative. One after another, deploying task forces would be taken through repetitive deployment preparation cycles using and reusing support services centralized ashore. To do this in the most efficient manner, the navy concentrated its training, supply, maintenance, and communications in as few hubs as possible—favoring bases already heavily concentrated—and San Diego's navy focus continued to grow.

For years after the end of World War II, ships were stowed in reserve at the 32nd Street Naval Station awaiting a return call to duty. Members of this "mothball fleet" included cruisers, destroyers, minesweepers, amphibious transports, and escort carriers in this photograph, taken *c.* 1956. (Naval Historical Center.)

This 1950s view of the radar room, also known as the Combat Information Center, aboard an *Essex*-class carrier shows the necessity for intensive team training in communications, plotting, team leadership, and tactical coordination. All these skills were honed through hours of practice at San Diego training sites, such as the Fleet Anti-submarine Training Center on Harbor Drive. (Naval Historical Center.)

At sea, a crewman standing watch updates the position of nearby surface contacts by writing backwards with a grease pencil on a status board on the bridge of aircraft carrier *Ranger* in 1964. Shipboard routine while underway remained unchanged for hundreds of years and is frequently repetitive. Watches generally lasted four hours at a time and occurred twice daily. (Naval Historical Center.)

Destroyer tender *Piedmont* is moored to a mid-channel buoy in the late 1950s with her small boat booms extended, awaiting the arrival of nesting destroyers. It was common practice during this time for destroyers to be moored in mid-stream rather than along piers at the naval station. Ferry *San Diego* or *Coronado* can be seen astern. (Maritime Museum of San Diego.)

Miramar's reputation as a center of high-performance jet operations was dramatically enhanced with the release of the movie *Top Gun* in 1986. The air station's celebrated Fightertown USA signboard reminded all on the F-14 flightline of Miramar's status, including this pilot and radar intercept officer from Fighter Squadron 194, who are climbing out of their aircraft after a mission in 1987. (U.S. Navy.)

A F-14A Tomcat from Fighter Squadron 124 flies over Miramar Naval Air Station (NAS) in May 1984. With the retirement of the F-14 by the late 1990s and with the realignment of West Coast navy and marine air bases, NAS Miramar was turned over to the marines in October 1997, completing a circle of ownership that began when the navy assumed control of Miramar from the marines in 1947. (U.S. Navy.)

The Tomahawk long-range land-attack cruise missile was built in San Diego and largely perfected in tests off the San Diego coast, including test firings from the battleship *New Jersey* off the coast of San Clemente Island in 1983. (Naval Historical Center.)

The stately, gull-winged Martin P5M Marlin seaplane was the last in a long line of navy flying boats in San Diego, stretching back to the first Curtiss A-1 of 1911. Seaplanes were a common site taxiing along San Diego Bay or flying over Coronado, as seen in this *c.* 1961 photograph. The last P5M flew out of San Diego in November 1967. (U.S. Navy.)

The FJ-4 Fury aircraft of Attack Squadron 126 break over North Island *c.* 1959. VA-126 was one of several squadrons based at Miramar in the training and replacement air-group role and evolved to provide "adversary" flight roles for TOPGUN (the popular name of the original U.S. Navy Fighter Weapons School) in the 1960s and 1970s. (Naval Historical Center.)

Recruits at the San Diego Naval Training Center exercised in many ways to develop esprit and a spirit of teamwork. Here they practice the traditional naval art of long boat, or cutter, racing in the center's boat channel in 1950. (National Archives.)

The San Diego Naval Training Center (NTC) ended its 75-year history in March 1997 and was returned to the city of San Diego by the navy. More than 1.75 million recruits and another million specialty school personnel graduated during its history. The NTC site (now referred to as Liberty Station) has been redeveloped as a location for mixed-use residential, retail, business, recreational, cultural, and educational activities. (National Archives.)

Evocative of a San Diego lifestyle of outdoor living and sunny days, Balboa Naval Hospital was particularly well-known for its attractive and airy central courtyard. Such attention to architectural detail and compatibility with the surrounding ambiance of Balboa Park also served to aid patient convalescence. Here the hospital staff is assembled for a change-of-command ceremony in April 1961. (U.S. Navy/Naval Facilities Engineering Command.)

Hospital ship *Repose* rounds Point Loma and returns to San Diego after an overseas deployment sporting a traditional "homeward bound" pennant held aloft by balloons astern. The homeward bound pennant is authorized by the navy in a long-standing tradition to recognize lengthy deployments of nine months or more and is designed to be proportional to the number of people aboard. The *Repose* served lengthy deployments in both Korea and Vietnam, providing remote hospital services. (National Archives.)

A primary spur to the San Diego economy, the navy's presence has helped nurture a host of waterfront industries that support its repair, maintenance, and supply requirements. National Steel and Shipbuilding Company (NASSCO), established in 1905 as a foundry, is one of several shipbuilding concerns on San Diego Bay and has a long history of warship construction. Above, Vice Adm. Edwin Hooper, third from left; Mrs. Edwin Hooper (ship's sponsor), fourth from left; Mrs. Edwin Hooper Jr. (matron of honor), third from right; and other members of the launching party for the USS *Racine* (LST-1191) gather for a picture before launch on August 15, 1970. At left, the new auxiliary cargo/ammunition ship *Alan Shepard* slides into San Diego Bay in 2006. (Naval Historical Center, U.S. Navy.)

Carrier operations are a hectic, danger-filled choreography involving dozens of flight deck crewmen that demands the sharpest edge of training and readiness for success. The Southern California fleet operating areas off San Diego span hundreds of square miles where this training has been constantly in motion from the very first aircraft carrier, *Langley*, to the most modern nuclear-powered carrier. Here, in the early days of the cold war, Air Group 11 aircraft await the signal to launch aboard carrier *Valley Forge* in 1948. Note the busy aircrewmen working on deck within inches of rapidly spinning propellers. (National Archives.)

Not all of the training the navy conducts in the fleet operating areas off the San Diego coast is easygoing with perfect Southern California weather, as longtime Pacific Fleet oiler *Mispillion* proves in this *c.* 1958 picture of operations in the "tanker lane" off San Clemente Island. (Naval Historical Center.)

Showing the civic spirit that has always defined the San Diego–U.S. Navy relationship, the city sponsored a "San Diego Salutes the Troops" parade in May 1991 to honor all veterans, especially those who were returning from Operation Desert Storm. Here a navy color guard leads a formation of sailors down Broadway. (U.S. Navy.)

Eight

The Carriers

No class of ship aligns with San Diego's ethos as well as the aircraft carrier. From the navy's first carrier, *Langley*, to Hollywood's treatment of TOPGUN, San Diego's imprint on carrier aviation is prominent and enduring. Nearly every carrier that has served in the U.S. Navy has visited San Diego at one time during its service—115 different carriers as estimated by the San Diego Air & Space Museum. North Island helped nurture the first advances in American naval aviation, and the efficient design of ready airfields, deep-water piers for aircraft carriers, and spectacular flying weather all helped underscore a continuing heritage of carrier aviation in San Diego.

Today an aircraft carrier at North Island is one of the first sights that visitors to San Diego see as they cruise down Interstate 5 or land at Lindbergh Field. And San Diego's largest single museum, fittingly, is the historic aircraft carrier *Midway*, berthed on the San Diego downtown waterfront.

When carrier *Langley* began experimenting with aircraft in exercises off the San Diego coast in 1925, she routinely carried six to eight aircraft. It was thought this number was the most that could be operated safely from her small deck. But by the time this photograph was taken in 1928, procedures had improved so that a full load of aircraft now numbered 34, dramatically increasing the offensive impact of a carrier strike and elevating the importance of carrier warfare in the eyes of naval officers. (National Archives.)

When carrier *Saratoga* gingerly entered the harbor and anchored close by Broadway Pier in November 1931, it marked a major milestone for carrier aviation in San Diego. The navy dredging programs beginning in 1912 had deepened the harbor sufficiently so that deep-draught ships, carriers, and battleships could now make it to the North Island turning basin, close to naval aircraft. (National Archives.)

Aircraft carrier *Ranger*, the first American ship designed and built from the keel up as an aircraft carrier, moored at San Diego shortly after her commissioning in April 1935. A victim of design compromises and limited experience in aircraft handling at sea, *Ranger*'s relatively small size limited her utility, and during the war, she served primarily in the Atlantic theater. (U.S. Navy.)

The *Essex* represented a new class of aircraft carriers that dramatically expanded the impact of naval aviation. *Essex* is shown here at the North Island quay in the early 1950s and *Essex*-class carriers would continue to be homeported in San Diego until the mid-1970s in both their attack carrier and ASW carrier designations. (San Diego Air & Space Museum.)

San Diego's carrier legacy is well documented in this December 1955 photograph, which captures six different carriers moored at North Island. Destroyer tenders and nests of destroyers were a common practice in the bay until the 1970s. (U.S. Navy.)

Independence represented the first generation of supercarriers to be introduced into the navy after World War II. Choosing San Diego as a home port for these large carriers (beginning in the early 1960s with *Kitty Hawk*) further solidified the city's importance to the navy and helped establish the "megaport" practice during the cold war, a practice where more and more naval services were concentrated in San Diego instead of at other West Coast ports.

Nuclear carriers like the *John C. Stennis* represent the present generation of carriers that operate from San Diego, the primary home port for aircraft carriers in the Pacific Fleet. (U.S. Navy.)

Between 1994 and 2001, the old carrier pier at North Island was replaced with two new quays to allow the homeporting of up to three nuclear-powered aircraft carriers. Here the nuclear-powered carrier *Nimitz* maneuvers alongside one of the new berths in 2001. (U.S. Navy.)

A young aviator and his wife sit in their car at the North Island quay in 1950, staring quietly at his waiting carrier before he boards for deployment to Korea. (National Archives.)

Nine

GATORS AND SEALS

During the first year of World War II, it became apparent that the U.S. Navy would need a robust capability to land troops ashore in order to move against the Japanese and to regain the European continent. Beyond a few exploratory landing exercises in the 1930s, the art and science of amphibious warfare was immature and untested. Whole new classes of specialized ships and landing craft would have to be designed and tested, tactics would have to be perfected, and communications and coordination honed. The navy, in large part, would be starting from scratch to develop these capabilities, and just as important, a new base was built in Coronado to centrally manage all amphibious activity for the Pacific Fleet.

One portion of amphibious doctrine called for reconnoitering beaches and clearing them of enemy obstructions ahead of the main waves of landing craft. This mission fell to a new elite corps of tactical swimmers and scuba divers that the navy gathered into Underwater Demolition Teams (UDT). Training for UDT also found its way to Coronado to integrate with other amphibious efforts. The navy's special warfare capability, manifested today in the navy's SEAL (Sea, Air, Land) teams, grew from this foundation, and the teams still train and operate from Coronado beaches.

By December 1943, a crash program begun only six months previously had created the Naval Amphibious Training Base by drawing fill from San Diego Bay. In 1944 and 1945, approximately 500 German prisoners of war were assigned to the base to complete the buildings for landing-force training. (National Archives.)

Over the years, San Clemente Island has provided navy men with a secure and convenient site for a wide spectrum of missions, from the testing of the latest weaponry to sophisticated and demanding training. Here in January 1944, *LST-486* practices beaching with a contingent of navy SeaBees (construction engineers) who must help land a load of trucks on anything other than a hospitable beach. (National Archives.)

Marines from the 1st Marines drive their amphibian assault vehicles from navy amphibious ships offshore near San Clemente Island to exercise modern amphibious warfare techniques. (U.S. Navy.)

Coronado's Silver Strand is almost entirely contained within a naval reservation, and it has long been favored with perfect conditions for amphibious training, whether in ocean surf conditions or in protected bay waters. These mechanized landing craft approach the beach in June 1982 to practice the all-important coordination of small boats in a landing wave. (U.S. Navy.)

Tank Landing Ships (LST) were invented during World War II to drive up on the beach and disgorge tanks, trucks, and bulky cargo through two large bow doors. The *Newport*-class LST *San Bernardino* takes this concept one step further by lowering a ramp onto a floating causeway off the Silver Strand in this *c.* 1970 photograph. (U.S. Navy.)

Amphibious ships of today are larger and more complex than their World War II counterparts but still make San Diego home. Here family and friends wave goodbye to loved ones aboard the amphibious assault ship *Peleliu* as she sails from the naval station for a six-month deployment in 2006. (U.S. Navy.)

Underwater Demolition Teams have had a presence in Coronado since World War II. A sign at the Naval Amphibious Base (above) in 1955 points to their wartime legacy. Below frogmen practice their patented high-speed boat pickups off a target beach nearby. (U.S. Navy/ Naval Special Warfare Command [NSWC].)

Underwater Demolition Teams (UDT) began experimenting with long-range helicopter insertion and extraction of frogmen as early as 1951, and that training continues today. A Sikorsky HUS-1 helicopter from Ream Field's HU-1 helicopter squadron practices with UDT members in San Diego Bay *c.* 1956 to help perfect this technique. (U.S. Navy/NSWC.)

The Mark V and other special operations craft are frequently seen training in San Diego Bay and off the Silver Strand, and are used by navy SEALs and special warfare combatants–craft crewmen for insertion and extraction of special operations forces and for coastal patrol and interdiction. The Mark V can travel at over 40 knots with water-jet engines and has a draft of only 5 feet for shallow water operations. (U.S. Navy.)

Frogmen of Underwater Demolition Team No. 1 demonstrate a Lambertson Amphibious Respiratory Unit in 1951, with standard UDT rubber boats in the background. After World War II, some 30 UDT teams were reduced to two in Coronado (UDT-1 and UDT-3) and two on the East Coast. UDT-5 was sent to Coronado in 1952 in response to Korean War requirements. (U.S. Navy/NSWC.)

Members of SEAL Team No. 5 carry a raft into the water before a night-combat swimming exercise off the Silver Strand. Much of the routine training for SEALs is conducted in Coronado and offshore at San Clemente Island. The Naval Amphibious Base is the home for all West Coast SEAL teams. (U.S. Navy.)

"The only easy day was yesterday" is a famous Basic Underwater Demolition/SEAL (BUDS) motto that is in daily practice at the Naval Special Warfare Center in Coronado. Formal training for new SEALs has been conducted at Coronado since 1950. BUDS training spans a rigorous yearlong regimen that includes training from basic to specialized efforts with teams. The first four weeks of physical and mental conditioning culminates in a demanding "Hell Week," a rugged test of physical, emotional, and mental motivation across nonstop ordeals from obstacle courses to group strength tests. (U.S. Navy.)

Ten

TODAY'S NAVY

Today San Diego stands at center stage of nearly every facet of American naval operations. It is the nerve center of a fleet that spans the globe and is a world leader in the maritime specialties of training, tactics, and technology. San Diego's sailors can be found on assignment throughout the Pacific. San Diego naval medical personnel are in combat zones in the Middle East. San Diego pilots fly daily missions against global terrorism. San Diego naval scientists test cutting-edge concepts in computers and communications.

A hundred thousand active-duty sailors and their families consider San Diego home, as do an even greater number of navy retirees, many of whom enjoy the city's charms after being introduced to San Diego through the happenstance of navy orders. The navy ranks as the largest single contributor to the San Diego economy, and whole industries thrive in San Diego to support its needs.

As important, within the navy it seems there has always been a certain San Diego mystique. A great percentage of navy men and women hold San Diego in high regard, many treasure the chance for San Diego assignments, and others seek the balance between professional enhancement and family stability that back-to-back assignments within the broad San Diego naval concentration can provide.

The navy in San Diego today is active, complex, and stimulating. Its presence touches hundreds of thousands of San Diegans every hour. And its stability and relevance argues for many more years of important contributions to San Diego's success.

The highlight of the annual Fleet Week celebration in San Diego has always been the Sea and Air Parade of ships into the harbor, which attracts over 100,000 spectators. Begun in 1997 as a means for the community to honor the men and women of the sea services, Fleet Week festivities involve parades, banquets, balls, scholarship awards, golf tourneys, ship tours, and the annual Miramar Air Show. (U.S. Navy.)

The navy's Blue Angels thrill the annual throng of spectators during the three days of the Miramar Air Show. Always a crowd favorite with aerobatics, aerial flybys, up-close aircraft displays, and free admission, the late summer air show has evolved its own San Diego-unique culture of family picnics on the sunny flight line and standstill traffic on Interstate 15. (U.S. Navy.)

Hospital ship *Mercy* is homeported in San Diego and is staffed largely by local medical care professionals from San Diego's Naval Medical Center. Routinely kept in a reserve status at the naval station, she is a striking presence on the San Diego waterfront. *Mercy* is activated periodically for humanitarian missions to the western Pacific and southwest Asia, such as this deployment in 2006. (U.S. Navy.)

Interaction between the navy and the community occurs in many forms, from free band concerts, charitable giving, and honor guards at civic events, to high-impact, red-white-and-blue displays such as the one seen here at a Chargers game in August 2004. The navy is one of several things that make San Diego unique, and the harmony that city and navy share is notable for its depth. (U.S. Navy.)

Beginning in 2001, the navy significantly upgraded its harbor security and force protection in San Diego Bay. Where once it was commonplace for tour boats and recreational sailors to come close enough to touch a ship's hull, warships (such as the nuclear-powered carrier *Nimitz*) are now surrounded by layers of security, including large floating-barrier booms and security patrol boats. (U.S. Navy.)

San Diego sailors wave from the deck of the amphibious dock landing ship *Comstock* in 2006. (U.S. Navy.)

The nuclear-powered aircraft carrier *Ronald Reagan* twists in the deep waters of the North Island turning basin with the help of a nearby tug in 2004 and heads to sea while her crew mans the rails. (U.S. Navy.)

For over 80 years, the navy either outwardly opposed or complicated efforts to build a bridge across San Diego Bay—disputes that spanned nine separate bridge or tunnel planning attempts. In 1964, the navy removed its objections as long as the central span provided clearance for the navy's largest ships, such as the amphibious assault ship *Belleau Wood*, seen here transiting under the bridge in 2005. Unfortunately this prerequisite (and the alignment of San Diego's freeways) required that the bridge be built south of downtown, necessitating that North Island's heavy rush-hour traffic traverse the entire breadth of residential Coronado. (U.S. Navy.)

Many a child remembers the thrilling return of a father or mother from a lengthy overseas deployment, such as this return of aircraft from Sea Control Squadron 33 and aircraft carrier *Carl Vinson* in 2005. Lt. Don Rogers reunites with his son on the North Island flightline. (U.S. Navy.)

A San Diego scene that has been reenacted hundreds of thousands of times over the years: Excited family and friends gaze toward an approaching ship to welcome home loved ones from war or extended deployments. Here family members gather at a naval station pier in 2003 to welcome home amphibious assault ship *Tarawa* from a 2003 deployment. (U.S. Navy.)

In 2004, the *Midway* aircraft carrier museum opened on the San Diego waterfront. Originally commissioned in 1945 (the largest ship in the world at the time), *Midway*'s service spanned 47 years. Today she provides a comprehensive interpretation of both carrier aviation and the San Diego–U.S. Navy heritage to over a million visitors a year who can tour her hangar and flight decks, the admiral's quarters, the Combat Information Center, and other spaces, and also view a world-class collection of restored and period aircraft. Above, *Midway* steams through heavy seas in 1949, and, below, she sits pier-side at her current site in downtown San Diego. (Above, U.S. Navy; below, San Diego Aircraft Carrier Museum.)

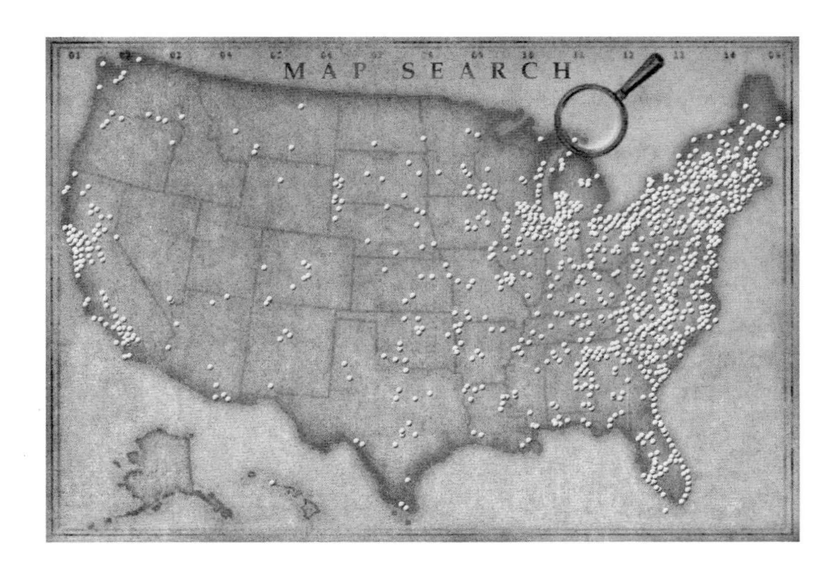
MAP SEARCH